Emu Companion

The Ultimate Guide to Raising and Caring for Emus as Pets: Everything You Need to Know About Feeding, Health, Behavior, and lots More!

Michael S. Rowen

Emu companion

Copyright © 2024 by Michael S. Rowen

All rights reserved. No part of this publication may be reproduced, distributed, or transmitted in any form or by any means, including photocopying, recording, or other electronic or mechanical methods, without the prior written permission of the publisher, except in the case of brief quotations embodied in critical reviews and certain other noncommercial uses permitted by copyright

Emu companion

TABLE OF CONTENTS

INTRODUCTION 6
    Why Choose an Emu as a Pet? 6
    Brief History and Background of Emus 8
    Understanding the Commitment: Responsibilities of Emu Ownership 8

CHAPTER 1 11
    GETTING STARTED WITH EMUS 11
    1.1 Legal and Practical Considerations 11
    1.2 Choosing the Right Emu: Breeds and Varieties .. 12
    1.3 Setting Up the Perfect Enclosure 14

CHAPTER 2 17
    FEEDING AND NUTRITION 17
    2.1 Emu Dietary Needs and Requirements 17
    2.2 Creating Balanced and Nutritious Meals 18
    2.3 Handling Special Dietary Considerations 20

CHAPTER 3 22
    HEALTH AND WELLNESS 22
    3.1 Common Health Issues in Emus 22
    3.2 Preventative Care: Vaccinations and Parasite Control 23
    3.3 Recognizing Signs of Illness and When to Call the Vet 25

CHAPTER 4 28
    EMU BEHAVIOR AND TRAINING 28
    4.1 Understanding Emu Behavior: Instincts and Social Dynamics 28

Emu companion

    4.2 Basic Training Techniques.................................29

    4.3 Enrichment Activities for Mental Stimulation....31

CHAPTER 5..**34**

  EMU REPRODUCTION AND BREEDING............................**34**

    5.1 Reproductive Biology of Emus..........................34

    5.2 Breeding Considerations and Challenges...........35

    5.3 Caring for Emu Chicks......................................37

CHAPTER 6..**40**

  EMUS IN YOUR DAILY LIFE.......................................**40**

    6.1 Integrating Emus into Your Household..............40

    6.2 Bonding and Building Trust...............................42

    6.3 Emu-Safe Enrichment and Entertainment Ideas.43

CHAPTER 7..**46**

  BEYOND THE BASICS: ADVANCED TOPICS.....................**46**

    7.1 Advanced Training and Agility..........................46

    7.2 Emu Shows and Competitions...........................48

    7.3 Conservation Efforts and the Future of Emus.....50

CHAPTER 8..**53**

  HOUSING AND HOUSING SETUP FOR EMUS....................**53**

CHAPTER 9..**59**

  GROOMING AND CARE..**59**

    REFLECTING ON YOUR JOURNEY WITH EMUS
....66

    Final Tips for Successful Emu Ownership................67

    Resources and Further Reading..............................68

    Final Thoughts..69

Emu companion

FREQUENTLY ASKED QUESTIONS (FAQS) ABOUT EMUS AS PETS 71

Emu companion

INTRODUCTION

Welcome to the World of Emus

Welcome to the enchanting world of emus, where the extraordinary meets the everyday! If you've ever dreamed of sharing your life with an animal that's as fascinating as it is affectionate, you've come to the right place. Emus, the second-largest birds in the world, offer a unique blend of companionship, intelligence, and personality. Their striking presence and engaging behaviors make them not just pets, but also wonderful companions that can enhance your life in countless ways.

Owning an emu is a journey filled with discovery and joy. As you delve deeper into this book, you will learn how to navigate the nuances of emu ownership, from their care and feeding to their social needs and health considerations. The path to emu companionship is as exciting as it is rewarding, and this guide will help you prepare for every step of the way.

Why Choose an Emu as a Pet?

Emus may not be the first animals that come to mind when you think of pets, but they bring an array of advantages that make them worth considering. Here are some compelling reasons to choose an emu as a pet:

1. **Unique Companionship:** Emus are highly social creatures that can form deep bonds with their caretakers. They are known for their playful and curious nature, often following their owners around, showcasing personalities that are both quirky and endearing. Unlike traditional pets, emus can engage

you in their world, making for an entertaining and interactive experience.
2. **Low Maintenance:** In contrast to more common pets, emus require less daily upkeep. They don't need walks, and their grooming needs are minimal. This makes them a suitable option for busy individuals or families seeking a pet that adds joy without a hefty time commitment.
3. **Educational Value:** Owning an emu provides an incredible educational opportunity. Their unique biology and behaviors can spark interest in avian science and conservation. This can be particularly enriching for children, fostering curiosity and respect for nature.
4. **Contribution to Conservation:** By caring for emus, you are contributing to the preservation of this fascinating species. Many emus are kept in captivity as part of breeding programs aimed at maintaining genetic diversity and preventing extinction, giving you a sense of purpose in your ownership.
5. **Environmental Benefits:** Emus can help maintain the ecosystem of your garden or yard. Their foraging habits assist in controlling insect populations, making them beneficial allies in maintaining a balanced environment.

By choosing an emu, you invite a unique and engaging companion into your life, one that can enrich your experiences and broaden your understanding of wildlife.

Emu companion

Brief History and Background of Emus

Emus (*Dromaius novaehollandiae*) are native to Australia, where they roam freely across diverse landscapes, from grasslands to forests. As members of the ratite family, which includes ostriches, rheas, and kiwis, emus have a long evolutionary history that dates back millions of years.

Their evolutionary traits, such as flightlessness, have allowed them to adapt remarkably well to their environment. With no significant natural predators in their native habitats, emus have thrived, developing characteristics that make them both resilient and adaptable.

Culturally, emus hold a significant place in Indigenous Australian societies. They are featured in various Dreamtime stories, highlighting their importance in the cultural and spiritual landscape. Historically, emus provided vital resources, including meat and feathers, playing a crucial role in the livelihoods of Indigenous communities.

While emus are not currently classified as endangered, they have faced challenges due to habitat destruction and hunting in the past. Today, conservation efforts are helping to stabilize their populations, allowing these magnificent birds to continue thriving in the wild and in human care.

Understanding the Commitment: Responsibilities of Emu Ownership

Before welcoming an emu into your life, it's essential to understand the responsibilities that come with ownership. Emus are not typical pets; they require a level of commitment that reflects their unique needs and characteristics.

Emu companion

1. **Long-Term Commitment:** Emus can live for 15 years or longer, making them a significant long-term commitment. Prospective owners must be prepared for the financial, emotional, and physical responsibilities involved in their care.
2. **Space and Environment:** Emus need ample space to roam and thrive. A suitable environment includes secure fencing to protect them from predators and a comfortable shelter to shield them from harsh weather conditions.
3. **Dietary Requirements:** A balanced diet is essential for maintaining an emu's health. Owners must be knowledgeable about their nutritional needs and be prepared to provide a varied diet that includes grains, fruits, and vegetables.
4. **Health Care:** Regular veterinary care is crucial for the well-being of your emu. This includes routine check-ups, vaccinations, and emergency care when necessary. Being vigilant about their health is paramount for ensuring a happy and thriving pet.
5. **Social Interaction:** As social creatures, emus thrive on interaction and companionship. Owners must dedicate time to bond with their emu and provide enrichment activities to keep them mentally stimulated and engaged.
6. **Legal Considerations:** Understanding local laws and regulations regarding emu ownership is vital. Ensure compliance with zoning laws and permits, which may vary from one area to another.

Emu ownership is a rewarding journey that comes with its challenges. By understanding these responsibilities, you can ensure a fulfilling and enriching experience for both you and your emu.

Emu companion

This introduction sets the stage for the journey ahead, offering insights into the remarkable world of emus, the joys of ownership, and the responsibilities that come with caring for these extraordinary birds. As you dive deeper into this book, you will gain the knowledge and tools necessary to embrace the unique experience of being an emu owner, creating a bond that will enhance both your lives.

CHAPTER 1

GETTING STARTED WITH EMUS

1.1 Legal and Practical Considerations

Before you embark on your journey of emu ownership, it is essential to navigate the legal landscape surrounding the ownership of these fascinating birds. Emus are classified as exotic pets in many regions, and understanding the legal requirements is crucial for ensuring a responsible and lawful ownership experience.

1. **Research Local Laws**
   The first step in preparing for emu ownership is to research your local laws regarding exotic animals. Regulations can vary significantly from one jurisdiction to another. Some areas classify emus as livestock, requiring different permits and adherence to specific agricultural guidelines, while others may treat them as exotic pets with their own set of rules.

   Check with your local animal control agency or agricultural department to determine what is required in your area. This may include permits for keeping emus, as well as specific zoning laws that govern the type of animals you can have on your property. Failing to comply with these regulations can lead to fines or the forced removal of your emu.

2. **Assess Zoning Regulations**
   Beyond state and local laws, your homeowners' association (HOA) or community guidelines may have additional restrictions. Some neighborhoods have rules prohibiting the keeping of livestock or

exotic pets, so it is essential to ensure that you are in compliance to avoid potential disputes with neighbors.

3. **Insurance Considerations**
   Owning an emu may also impact your homeowner's insurance policy. It's wise to consult with your insurance provider to discuss whether keeping an emu is permissible under your current policy. Some insurers may require additional coverage for exotic pets, while others may have restrictions that you need to consider. This step will help protect you financially in the event of any incidents involving your emu.

4. **Veterinary Care**
   Finding a qualified veterinarian experienced in avian care is critical for maintaining the health of your emu. Not all veterinarians are equipped to handle exotic birds, so take the time to locate a practice that specializes in avian medicine. Regular health check-ups, vaccinations, and preventative care are essential to ensure that your emu remains healthy and free from diseases.

By understanding and addressing these legal and practical considerations, you will create a secure foundation for your emu ownership journey, allowing for a harmonious and responsible relationship with your new companion.

1.2 Choosing the Right Emu: Breeds and Varieties

Once you have a solid understanding of the legal requirements, the next step is to select the right emu for your home. Emus come in various breeds and varieties, each with unique characteristics and temperaments that can influence your experience as an owner.

Emu companion

1. **Standard Emus**
   The most common type of emu kept as a pet is the standard emu (*Dromaius novaehollandiae*). Typically standing between 5 to 6 feet tall and weighing between 60 to 100 pounds, these emus are known for their friendly and sociable nature. They are well-adapted to human interaction, often displaying curious behaviors that make them engaging companions.
2. **Miniature Emus**
   For those who have limited space or prefer a smaller bird, miniature emus are an excellent option. These birds typically stand around 4 to 5 feet tall, making them more manageable for smaller yards or properties. While still requiring plenty of room to roam and explore, their size can make them a more feasible option for potential owners with space constraints.
3. **Color Varieties**
   Emus also exhibit various feather colorations, including shades of blue, gray, and brown. These color variations can enhance the aesthetic appeal of your flock and add a unique charm to your emu ownership experience. While color may not significantly impact the care requirements of the bird, it can influence your personal preference when selecting your emu.
4. **Health and Temperament**
   When choosing an emu, it's crucial to consider not just appearance but also health and temperament. Look for emus that display signs of good health, such as bright, clear eyes and shiny feathers. Active and inquisitive behavior is also an indicator of a healthy emu.

It is highly recommended to purchase your emu from a reputable breeder or rescue organization, as these sources are more likely to provide well-socialized birds that have been properly cared for. A well-chosen emu will contribute to a fulfilling and enriching companionship.

Carefully selecting the right emu, you ensure that your new companion is well-suited to your lifestyle and environment, setting the stage for a rewarding relationship.

1.3 Setting Up the Perfect Enclosure

Creating a safe and comfortable living environment is essential for the health and happiness of your emu. A well-designed enclosure will allow your emu to thrive and express its natural behaviors.

1. **Space Requirements**
   Emus are active birds that require ample space to roam and forage. A general guideline is to provide at least 1000 square feet of outdoor space per emu. The more space you can offer, the better, as this allows them to engage in their natural behaviors and socialize with you and other emus.
2. **Fencing**
   Security is paramount when housing emus. A robust and secure fence is essential to keep your emu safe from predators and to prevent escape. Fencing should be at least 6 feet tall and constructed from strong materials, such as chain-link or wooden boards. Emus are naturally curious creatures, and they can jump or climb over inadequate barriers, so it is essential to ensure the enclosure is secure.
3. **Shelter**
   In addition to space, emus need a sheltered area to

protect them from the elements. This shelter can be a simple structure that provides shade from the sun and protection from rain or wind. Ensure the shelter is well-ventilated to maintain a comfortable environment. Providing bedding materials, such as straw or wood shavings, can help keep the area clean and comfortable.

4. **Enrichment Features**
Emus are intelligent animals that require mental stimulation to prevent boredom. Incorporating enrichment features into their enclosure can enhance their quality of life. This might include logs for perching, water features for bathing, and plants for foraging. These elements encourage natural behaviors, allowing your emu to engage in exploration and play.

5. **Cleanliness**
Regular cleaning of the enclosure is vital to maintaining a healthy living environment. Establishing a routine for cleaning and sanitizing the area will help reduce the risk of disease and ensure a safe habitat for your emu. Monitor waste levels and remove droppings regularly to keep the area clean and odor-free.

However, setting up the perfect enclosure, you create a safe and stimulating environment for your emu, allowing it to thrive and enjoy its new home.

In this chapter, we have explored the essential steps to getting started with emus, focusing on the legal and practical considerations, choosing the right emu, and setting up an ideal living environment. With careful preparation and thoughtful decision-making, you can create a nurturing space that supports the health and happiness of your emu,

Emu companion

paving the way for a fulfilling companionship that can last for years to come.

CHAPTER 2

FEEDING AND NUTRITION

2.1 Emu Dietary Needs and Requirements

Understanding the dietary needs of emus is crucial for ensuring their health and well-being. As omnivores, emus require a diverse diet that provides them with the necessary nutrients to thrive. Here's a breakdown of their dietary needs:

1. **Basic Nutritional Components**
   Emus require a balanced diet that includes the following nutritional components:
   - **Proteins:** Essential for growth, development, and overall health, proteins can be sourced from various foods, including grains, legumes, and commercial emu feeds.
   - **Carbohydrates:** These provide energy, and emus primarily derive carbohydrates from grains such as corn, wheat, and oats.
   - **Fats:** Healthy fats are vital for maintaining skin and feather health. Sources include seeds, nuts, and commercially available poultry fat.
   - **Vitamins and Minerals:** Emus need a range of vitamins and minerals for optimal health. Key nutrients include calcium for bone development, phosphorus for cellular function, and vitamins A, D, and E for immune health and reproductive function.
2. **Water Intake**
   Access to fresh, clean water is critical for emus.

These birds can drink significant amounts of water, especially in hot weather. It's essential to provide a constant supply of fresh water to keep them hydrated and promote overall health. Water should be kept clean and changed regularly to prevent contamination.
3. **Foraging Behavior**
In their natural habitat, emus are foragers, constantly searching for food. They enjoy exploring their environment, which allows them to engage in natural behaviors. Providing opportunities for foraging not only keeps them mentally stimulated but also supports their dietary needs. This can include scattering feed or using foraging toys that encourage them to search for hidden treats.

Understanding the dietary needs and requirements of emus, you can create a feeding plan that supports their health, enhances their natural behaviors, and contributes to their overall well-being.

2.2 Creating Balanced and Nutritious Meals

Creating balanced and nutritious meals for your emu is essential for its growth and health. Here's how to ensure your emu's diet is well-rounded and fulfilling:

1. **Commercial Emu Feeds**
Many pet stores and agricultural suppliers offer specially formulated commercial emu feeds. These feeds are designed to meet the specific nutritional needs of emus and often contain a balanced blend of grains, proteins, vitamins, and minerals. When selecting a commercial feed, choose a reputable brand that is appropriate for the age and activity level of your emu.

2. **Supplementing with Fresh Foods**
   In addition to commercial feeds, supplementing your emu's diet with fresh foods can enhance nutrition and provide variety. Safe options include:
   - **Fruits:** Emus enjoy fruits like apples, bananas, berries, and melons. These provide essential vitamins and hydration but should be given in moderation due to their sugar content.
   - **Vegetables:** Leafy greens (like spinach and kale), carrots, and squash are excellent choices. These provide fiber, vitamins, and minerals.
   - **Grains and Seeds:** Incorporate grains such as oats, barley, and sunflower seeds to enhance the diet and provide healthy fats.
3. **Avoiding Toxic Foods**
   While many foods are safe for emus, some are toxic and should be avoided. Foods such as chocolate, avocados, and onions can be harmful to birds and should never be included in their diet. Always research or consult with a veterinarian about any new foods you wish to introduce.
4. **Feeding Frequency and Portions**
   Emus typically eat twice a day, but the feeding frequency may vary based on their age, size, and activity level. Young emus may require more frequent feeding due to their growth needs. Monitor your emu's weight and adjust portion sizes accordingly to prevent obesity. The diet should be tailored to your emu's individual needs, taking into account its age, size, and health status.
5. **Monitoring Health**
   Regularly monitoring your emu's health is crucial. Keep an eye on its weight, plumage, and overall demeanor. Any sudden changes in appetite or

behavior may signal dietary issues or health problems that should be addressed with a veterinarian.

Creating balanced and nutritious meals, you ensure that your emu receives the essential nutrients it needs for growth, development, and overall health, contributing to a long and vibrant life.

2.3 Handling Special Dietary Considerations

As with any pet, some emus may have specific dietary needs or health concerns that require special attention. Here are key considerations for managing these situations:

1. **Age-Specific Diets**
   The dietary needs of emus can change with age. Young emus (chicks) require a diet higher in protein to support rapid growth, while adults may need a more balanced approach to maintain weight and health. Pay attention to the specific nutritional requirements at different life stages, and adjust their diet accordingly.
2. **Health Conditions**
   Emus may develop health issues that necessitate dietary adjustments. For example, emus with metabolic disorders might require a low-calcium diet, while those with digestive issues may benefit from a higher fiber intake. If you notice any health concerns, consult with a veterinarian to create a tailored feeding plan that addresses these specific needs.
3. **Weight Management**
   Obesity can be a concern in captive emus, especially if they are fed high-calorie treats or are not provided with enough space to roam and forage.

Regularly monitor your emu's weight and adjust its diet and exercise routine to maintain a healthy weight. Offering a variety of low-calorie vegetables can help keep them satisfied without contributing to weight gain.

4. **Seasonal Diet Changes**
   Seasonal changes can also affect an emu's dietary needs. During warmer months, your emu may require more water and fresh fruits to stay hydrated. In colder months, consider providing extra calories to help maintain body heat. Adjusting the diet based on seasonal factors ensures that your emu remains comfortable and healthy throughout the year.

5. **Consulting a veterinarian**
   Whenever you encounter a dietary concern or health issue, consulting with a veterinarian experienced in avian care is vital. They can provide tailored advice and dietary recommendations based on your emu's specific health needs, ensuring it receives the best possible care.

In addressing special dietary considerations, you can provide your emu with a personalized feeding plan that supports its health and well-being, allowing it to thrive in its environment.

In this chapter, we have delved into the essential aspects of feeding and nutrition for emus, including their dietary needs, how to create balanced meals, and how to handle special dietary considerations. By understanding these fundamental principles, you can ensure that your emu receives the necessary nutrition to lead a healthy and vibrant life, contributing to the joy of your companionship for years to come.

CHAPTER 3

HEALTH AND WELLNESS

3.1 Common Health Issues in Emus

Ensuring the health and well-being of your emu requires an understanding of common health issues that may arise. By being aware of these conditions, you can take proactive measures to prevent and address them effectively.

1. **Respiratory Infections**
   Respiratory infections are among the most common health issues faced by emus. Symptoms can include coughing, nasal discharge, and labored breathing. These infections are often caused by bacteria or viruses and can be exacerbated by poor housing conditions, such as inadequate ventilation or high humidity. Maintaining a clean, dry, and well-ventilated living environment is crucial for preventing respiratory problems.
2. **Foot and Leg Problems**
   Emus are prone to various foot and leg issues, including bumblefoot (a bacterial infection) and tendon injuries. Bumblefoot manifests as swelling and sores on the feet, typically caused by rough surfaces or unsanitary conditions. To minimize the risk of foot problems, ensure your emu's enclosure has soft, clean bedding and avoid sharp or abrasive surfaces. Regularly inspect your emu's feet for any signs of swelling or lesions.
3. **Digestive Disorders**
   Digestive issues can occur due to dietary imbalances or abrupt changes in diet. Common symptoms include diarrhea, bloating, and lethargy.

Emus require a balanced diet that includes fiber-rich foods to support healthy digestion. If you notice any digestive problems, consult a veterinarian promptly to identify the cause and implement appropriate dietary changes.
4. **Obesity**
Obesity is a growing concern in captive emus, primarily due to overfeeding or lack of exercise. Excess weight can lead to various health complications, including joint problems and metabolic disorders. Regular monitoring of your emu's weight and activity levels, along with providing a balanced diet, can help prevent obesity.
5. **Reproductive Issues**
Female emus can experience reproductive issues, such as egg binding, where an egg becomes trapped in the oviduct. Symptoms may include lethargy, straining, and lack of appetite. If you suspect your emu is experiencing reproductive problems, it is essential to consult a veterinarian for immediate evaluation and treatment.

Being vigilant and proactive about these common health issues, you can enhance your emu's overall wellness and quality of life.

3.2 Preventative Care: Vaccinations and Parasite Control

Preventative care is essential in ensuring your emu remains healthy and free from disease. This section will cover the importance of vaccinations and parasite control.

1. **Vaccinations**
Vaccinations play a vital role in protecting your emu from infectious diseases. It is essential to

Emu companion

consult with a veterinarian about the recommended vaccination schedule, which may vary depending on your location and the specific health risks in your area. Common vaccines for emus include:
- **Newcastle Disease Vaccine:** This vaccination protects against a viral disease that can cause respiratory and nervous system issues.
- **Avian Influenza Vaccine:** This vaccine is crucial in preventing a highly contagious viral infection that can lead to severe illness or death in birds.

Keeping accurate vaccination records is essential for monitoring your emu's health and ensuring timely vaccinations.

2. **Parasite Control**
Internal and external parasites can significantly impact your emu's health. Regular parasite control measures are essential to prevent infestations. Common parasites that may affect emus include:
   - **Internal Parasites:** Worms, such as roundworms and tapeworms, can cause digestive issues and weight loss. Regular fecal examinations by a veterinarian can help identify and treat any internal parasites.
   - **External Parasites:** Mites and lice can cause skin irritation and discomfort. Regularly inspect your emu for signs of external parasites and consult a veterinarian for appropriate treatment options, such as medicated powders or sprays.
3. **Regular Health Check-ups**
Establishing a routine of regular veterinary check-ups is an important aspect of preventative care.

These check-ups allow for early detection of potential health issues and provide an opportunity to discuss any concerns you may have regarding your emu's health and nutrition.

Implementing a comprehensive preventative care plan that includes vaccinations, parasite control, and regular veterinary visits, you can significantly reduce the risk of illness and ensure your emu leads a healthy life.

3.3 Recognizing Signs of Illness and When to Call the Vet

Being able to recognize the signs of illness in your emu is critical for prompt intervention and treatment. Emus are generally stoic creatures, often hiding their discomfort until conditions become severe. Here are key indicators of illness to watch for:

1. **Changes in Appetite and Drinking Habits**
   A sudden decrease in appetite or water intake can indicate health issues. If your emu shows signs of disinterest in food or water, monitor it closely and consult a veterinarian if the behavior persists for more than 24 hours.
2. **Behavioral Changes**
   Pay attention to any significant changes in your emu's behavior. Lethargy, increased aggression, or withdrawal from social interactions can signal underlying health problems. An emu that is not engaging in its typical activities may be in distress and should be evaluated by a veterinarian.
3. **Physical Symptoms**
   Look for any physical signs of illness, such as:
   - **Respiratory Issues:** Coughing, wheezing, nasal discharge, or labored breathing.

- **Digestive Disturbances:** Diarrhea, bloating, or any signs of discomfort in the abdominal area.
- **Skin Conditions:** Swelling, sores, or changes in feather condition.

Document any unusual physical symptoms and report them to your veterinarian, as this information will assist in diagnosis and treatment.

4. **When to Call the Vet**
   If you notice any concerning signs or symptoms in your emu, it is essential to consult with a veterinarian as soon as possible. Early intervention can often lead to more successful treatment outcomes. Additionally, if your emu experiences sudden changes in behavior, severe weight loss, or displays any severe symptoms (such as excessive bleeding or difficulty breathing), seek immediate veterinary attention.
5. **Establishing a Relationship with Your Veterinarian**
   Building a relationship with a veterinarian who specializes in avian care will ensure that you have access to the best possible guidance for your emu's health. They can provide tailored advice, regular health checks, and support in emergencies, contributing to a long and healthy life for your companion.

Being attentive to your emu's health and recognizing signs of illness, you can ensure timely intervention and treatment, promoting a healthy and fulfilling life for your emu.

This chapter explored essential aspects of health and wellness for emus, including common health issues, preventative care strategies such as vaccinations and parasite control, and the importance of recognizing signs of illness. By prioritizing your emu's health and wellness, you contribute significantly to its quality of life and ensure a long, happy companionship.

CHAPTER 4

EMU BEHAVIOR AND TRAINING

4.1 Understanding Emu Behavior: Instincts and Social Dynamics

To foster a healthy relationship with your emu, it is essential to understand its natural instincts and social behaviors. Emus are unique creatures with specific behavioral traits influenced by their biology and environmental factors.

1. **Instincts**
   Emus possess several instinctual behaviors that stem from their evolution as wild birds. Some of the primary instincts include:
   - **Foraging Behavior:** In the wild, emus spend a significant amount of time foraging for food. They are naturally curious and will explore their environment for edible items, including seeds, fruits, and insects. Providing opportunities for foraging is essential for their mental and physical well-being.
   - **Territoriality:** Emus can exhibit territorial behavior, especially during the breeding season. They may become protective of their space and may display aggressive behaviors toward perceived intruders. Understanding this instinct can help you manage interactions with other animals and people.
   - **Flocking Behavior:** Emus are social birds and often live in small groups. They communicate through various vocalizations and body language, reinforcing social bonds

within the group. Providing companionship through other emus can help satisfy their social needs and reduce stress.
2. **Social Dynamics**
Emus have complex social structures that can be influenced by their environment. Understanding these dynamics is crucial for proper interaction:
    - **Hierarchy:** Within a group, emus may establish a pecking order, with dominant individuals asserting control over resources like food and space. Observing your emu's interactions can help you identify its social position and manage any potential conflicts.
    - **Bonding:** Emus often form strong bonds with their caretakers and other emus. Positive interactions can lead to trust and a cooperative relationship. Engaging in regular, positive interactions will foster a strong bond between you and your emu.

By understanding emu behavior, you can create an environment that meets their instinctual needs, reducing stress and enhancing your relationship with these fascinating birds.

4.2 Basic Training Techniques

Training your emu can be an enjoyable and rewarding experience for both you and your bird. Utilizing positive reinforcement techniques can help establish desired behaviors and strengthen your bond.

1. **Establishing Trust**
Before you begin training, building trust is vital. Spend time observing and interacting with your emu in a calm and safe environment. Use gentle

movements and soft vocalizations to create a reassuring atmosphere. Offering treats, such as fruits or vegetables, can also help establish a positive association with you.
2. **Positive Reinforcement**
Positive reinforcement is one of the most effective training techniques. This method involves rewarding your emu for desired behaviors, which encourages them to repeat those behaviors. Here's how to implement it:
    - **Rewards:** Identify treats that your emu enjoys and use them as rewards during training sessions. These can include small pieces of fruits, vegetables, or commercial bird treats.
    - **Timing:** The reward should be given immediately after the desired behavior to create a clear connection between the action and the reward. This helps the emu understand what behavior is being reinforced.
    - **Consistency:** Be consistent in your training approach. Use the same cues and rewards for specific behaviors to help your emu learn effectively.
3. **Basic Commands**
Start with simple commands that are easy for your emu to understand. Some fundamental commands to consider include:
    - **Come:** Use a specific word or sound to encourage your emu to approach you. Reward them with a treat when they respond.
    - **Stop or Stay:** Teach your emu to stop or stay in place when commanded. Use a

verbal cue and reward them for remaining still.
- **Target Training:** Target training involves using a stick or your hand as a target. Encourage your emu to touch the target with its beak, rewarding them each time they successfully do so. This technique can be useful for guiding your emu during other training exercises.

4. **Patience and Persistence**
Training takes time and patience. Emus may not respond immediately, so remain calm and avoid frustration. Consistent, short training sessions are more effective than long, infrequent ones. Gradually increase the complexity of commands as your emu becomes more comfortable with training.

By employing basic training techniques based on trust and positive reinforcement, you can establish a strong bond with your emu while promoting desirable behaviors.

4.3 Enrichment Activities for Mental Stimulation

Mental stimulation is vital for emus, as it promotes overall well-being and prevents boredom-related behaviors. Engaging your emu in enrichment activities encourages natural behaviors and keeps them physically and mentally active.

1. **Foraging Opportunities**
Emulate their natural foraging behavior by providing scattered food or using foraging toys. Hide treats in different locations within their enclosure, encouraging your emu to search and explore. This activity not only provides mental

stimulation but also mimics their instinctual behavior.

2. **Obstacle Courses**
   Create a simple obstacle course in your emu's enclosure using safe, non-toxic materials. Include objects for your emu to navigate around or under, such as boxes, cones, or logs. This activity encourages physical activity and engages their problem-solving skills.

3. **Interactive Toys**
   Consider incorporating interactive toys designed for birds into your emu's environment. These toys can include puzzles that dispense treats when solved or toys that require manipulation. Regularly rotate the toys to maintain novelty and interest.

4. **Social Interaction**
   If possible, consider keeping more than one emu to provide companionship and social interaction. Socialization among emus encourages natural behaviors and reduces stress. If keeping multiple emus is not feasible, spend quality time engaging with your emu through play and interaction.

5. **Training Sessions as Enrichment**
   Incorporate training sessions as a form of mental enrichment. Teaching new commands or tricks provides both mental challenges and bonding opportunities. The positive interactions from training can enhance your emu's overall happiness and well-being.

By implementing various enrichment activities, you can ensure that your emu remains mentally stimulated and content, contributing to its overall health and happiness.

This chapter, explored the essential aspects of emu behavior and training, including their instincts and social dynamics, effective training techniques, and enrichment activities for mental stimulation. By understanding and addressing the behavioral needs of your emu, you can create a positive environment that fosters a strong bond and enhances the overall well-being of your avian companion.

CHAPTER 5

EMU REPRODUCTION AND BREEDING

5.1 Reproductive Biology of Emus

Understanding the reproductive biology of emus is essential for successful breeding and chick rearing. Emus exhibit unique reproductive characteristics that set them apart from many other bird species.

1. **Breeding Season**
   Emus typically breed in the spring and summer months, with peak activity depending on geographical location and climate. During this period, males and females engage in courtship behaviors that include vocalizations, displays, and physical interactions. Males will often initiate courtship by performing a series of elaborate movements, such as bowing and fluffing their feathers, to attract females.
2. **Mating Behavior**
   Emus are generally monogamous during the breeding season, forming pairs that may mate for several seasons. After a successful courtship, mating occurs, with the male and female engaging in cloacal contact, which facilitates sperm transfer. After mating, the female will begin to lay eggs, usually producing a clutch of 5 to 15 eggs, depending on her health and environmental conditions.
3. **Egg Characteristics**
   Emu eggs are large, typically weighing around 500 grams each, and have a distinct dark green or blue-

green color with a smooth texture. The eggs are laid in a nesting area, which the female prepares by creating a depression in the ground and lining it with vegetation. Emu eggs have a thick shell that provides protection to the developing embryo.

4. **Incubation Period**
   The incubation period for emu eggs lasts approximately 49 to 56 days. Unlike many bird species, male emus take on the primary responsibility for incubation, sitting on the eggs to provide warmth and protection. During this time, the female may leave the nest to forage and regain her strength. Males will turn the eggs regularly to ensure even development.

5. **Hatching Process**
   When the chicks are ready to hatch, they use an egg tooth, a small, hard structure on their beak, to break through the eggshell. Hatching usually occurs over a period of a few days, with the male remaining vigilant to protect the newly hatched chicks from potential threats.

Understanding these biological processes is crucial for any emu owner interested in breeding, as it provides the foundation for successful reproduction and chick development.

5.2 Breeding Considerations and Challenges

While breeding emus can be rewarding, it also presents various challenges that require careful consideration and planning.

1. **Selecting Breeding Stock**
   Choosing healthy and genetically sound breeding stock is vital for producing strong, healthy chicks.

When selecting emus for breeding, consider the following:
- **Health Status:** Ensure that both the male and female are in good health, free from diseases and parasites.
- **Genetic Diversity:** Aim to breed emus from different genetic lines to reduce the risk of genetic disorders and improve overall flock vitality.

2. **Breeding Environment**

Creating a suitable breeding environment is essential for successful reproduction. The breeding area should be spacious, secure, and provide ample opportunities for foraging and nesting. Additionally, ensuring the environment is free from stressors (such as loud noises or aggressive animals) is crucial for encouraging successful mating.

3. **Monitoring for Stress**

During the breeding season, monitor both male and female emus for signs of stress. Breeding can be physically demanding, especially for males who incubate the eggs. Ensure they have access to food, water, and shelter, and avoid excessive handling during this period to reduce stress.

4. **Potential Challenges**

Breeding emus can present challenges, including:
- **Egg Infertility:** Not all eggs laid will be fertile. Factors such as poor nutrition or stress can contribute to infertility. Conduct regular checks on the eggs and consult with a veterinarian if fertility issues persist.
- **Nest Abandonment:** In some cases, a female may abandon her nest or fail to return for incubation. This can occur due to stress or environmental factors. If this

Emu companion

happens, consider using an incubator to ensure the eggs hatch successfully.
- **Predation:** Newly laid eggs and chicks are vulnerable to predators. Implement security measures, such as fencing and monitoring, to protect the nest and the chicks.

By carefully considering these breeding aspects, you can increase the chances of successful reproduction and contribute to the overall health and vitality of your emu flock.

5.3 Caring for Emu Chicks

Once the emu chicks hatch, they require attentive care to thrive and grow into healthy adults. Proper management during this early life stage is crucial for their development.

1. **Immediate Post-Hatching Care**
   After hatching, the chicks will stay with the male parent, who will continue to protect and nurture them. It is essential to allow the male to bond with the chicks and ensure they remain with him during the early days. The father will guide the chicks to food and water, teaching them essential survival skills.
2. **Feeding Requirements**
   Emu chicks have specific nutritional needs to support their rapid growth and development:
   - **Starter Feed:** Provide a high-quality starter feed formulated for game birds or poultry, which contains the necessary protein and nutrients. Ensure the feed is fresh and clean.
   - **Fresh Water:** Ensure that clean, fresh water is always available for the chicks. They may

need guidance to learn how to drink from bowls or troughs.
- **Supplementation:** As they grow, gradually introduce fresh greens, fruits, and vegetables to their diet. This variety will help them develop healthy eating habits and provide essential vitamins and minerals.

3. **Monitoring Growth and Health**
Regularly monitor the growth and health of the chicks. Healthy emu chicks should exhibit consistent weight gain, be active and curious, and have bright, alert eyes. Look for any signs of illness, such as lethargy or poor appetite. If you notice any concerning symptoms, consult a veterinarian for guidance.

4. **Socialization and Interaction**
Emu chicks are social animals and thrive in groups. If possible, allow them to interact with other chicks to encourage social bonding and proper development. Regular interaction with humans can also help them become accustomed to handling and establish trust.

5. **Transitioning to Adult Care**
As emu chicks grow and reach maturity, gradually transition them to adult care. This includes adjusting their diet to a balanced adult formula and providing a larger enclosure that accommodates their growth. Monitor their behavior and interactions as they integrate into the adult flock.

Providing a comprehensive care for emu chicks during their early stages, you can foster their development and prepare them for a healthy, productive life.

In this chapter, we delved into the intricacies of emu reproduction and breeding, including their reproductive

Emu companion

biology, breeding considerations and challenges, and the essential care required for emu chicks. Understanding these aspects is vital for any aspiring emu breeder, ensuring a successful breeding experience and fostering healthy, vibrant emu populations.

CHAPTER 6
EMUS IN YOUR DAILY LIFE

Emus, with their striking size, curious personalities, and unique behaviors, can be fascinating additions to your household. However, integrating them into your daily life requires careful planning, mutual respect, and dedication to their needs. In this chapter, we will explore the steps to smoothly integrate emus into your living environment, establish a strong bond of trust, and ensure they are enriched with safe and stimulating activities.

6.1 Integrating Emus into Your Household

Bringing an emu into your household is an exciting decision, but it comes with significant responsibilities. Unlike smaller pets, emus require a different approach in terms of housing, care, and interaction.

1. **Emu-Specific Living Arrangements**
   Emus cannot be housed like traditional pets. They need ample outdoor space that mimics their natural environment:
   - **Space Requirements:** Emus are large, active birds that require plenty of space to roam, run, and forage. A minimum of half an acre per emu is recommended, with secure fencing to keep them safe from predators and prevent them from wandering off.
   - **Outdoor Enclosure:** Their enclosure should be durable and high enough to prevent escape, with sturdy fencing at least six feet tall. Emus can jump and sprint at high

speeds, so it's essential to create a secure perimeter.
- **Shelter:** Although emus are hardy and can tolerate a range of temperatures, providing a simple shelter for protection against extreme weather (heavy rain, snow, or excessive heat) is important. The shelter should be large enough for the emu to stand and move comfortably.

2. **Coexisting with Other Pets**
Emus can cohabitate with other animals, but it is crucial to introduce them carefully:
    - **Introducing Emus to Dogs and Cats:** Emus are naturally curious but can be easily startled. When introducing them to dogs or cats, ensure the meeting is done gradually and under supervision. It is best to introduce calm and non-aggressive pets first.
    - **Other Birds and Livestock:** If you have other birds or livestock, emus may integrate well depending on the species. However, care must be taken to avoid aggressive behaviors between animals, especially during feeding times.

3. **Household Dynamics**
Emus have a strong sense of independence and may not adapt well to confined indoor environments. They are better suited to outdoor living but can form bonds with humans through daily interaction. If you choose to have regular close contact with your emu, it's important to ensure that all household members are comfortable with their presence and aware of how to handle them.

Integrating an emu into your household requires planning, space, and understanding. They are majestic animals that bring immense joy, but their size and needs must be carefully considered.

6.2 Bonding and Building Trust

Building trust and forming a bond with your emu is essential for a positive and rewarding relationship. Trust is the foundation of any interaction with an emu, as they are sensitive to how they are treated and approached.

1. **Establishing Trust from the Beginning**
   When you first bring an emu into your life, it is crucial to give them time to acclimate to their new environment. Start by spending time near the emu enclosure without directly interacting, allowing the bird to become familiar with your presence.
   - **Gentle Approach:** Avoid sudden movements or loud noises, as emus can be easily startled. Approach them slowly, speak softly, and offer food treats to build positive associations.
   - **Respecting Boundaries:** Just like any other animal, emus need personal space. Don't force interaction—let them come to you when they feel comfortable.
2. **Daily Interaction for Bonding**
   Regular, positive interaction is key to bonding with your emu:
   - **Feeding Time as Bonding Time:** Feeding your emu by hand or being present during feeding time helps establish a strong connection. Offering healthy treats such as fruits and vegetables can also reinforce positive interactions.
   - **Touch and Tactile Interaction:** Emus are sensitive to touch, so gradually introducing

physical contact is important. Start with gentle strokes on their back or neck, always being mindful of their comfort level.
3. **Building a Routine**
Emus thrive on consistency. Establishing a daily routine helps build trust, as they become familiar with the predictable patterns of feeding, cleaning, and enrichment activities. The routine fosters a sense of security and reduces stress in the bird.
4. **Signs of Trust**
When an emu feels comfortable with you, they will display trust in various ways, such as approaching you on their own, allowing gentle touch, or following you around. Building a trusting relationship with your emu will lead to a deeper bond and more enjoyable interactions.

Trust is earned slowly with emus, and through patience and consistent positive reinforcement, you will develop a deep and meaningful connection with your bird.

6.3 Emu-Safe Enrichment and Entertainment Ideas

Emus are intelligent and active birds that need mental and physical stimulation to stay healthy and content. Providing enriching activities not only improves their quality of life but also strengthens your bond with them.

1. **Foraging and Food-Based Enrichment**
Emus are natural foragers, and you can replicate this behavior in a domestic environment:
    - **Scattering Feed:** Rather than offering food in a bowl, scatter it around their enclosure to encourage them to forage, which provides

Emu companion

both physical activity and mental stimulation.
- **Puzzle Feeders:** Using puzzle feeders or treat-dispensing toys can challenge your emu to figure out how to access their food, adding an extra layer of enrichment.
- **Variety in Diet:** Occasionally introduce new and safe food items, such as fruits, vegetables, and insects, to keep their diet exciting and engage their curiosity.

2. **Physical Activities**

Emus need plenty of exercise to stay fit and healthy:
- **Obstacle Courses:** Create simple obstacle courses in their enclosure using logs, boxes, and tires for them to navigate. This encourages them to explore and improves their coordination.
- **Running Space:** Since emus are built to run, ensure their enclosure provides enough space for them to sprint and exercise. This activity is essential for their well-being.
- **Interactive Play:** Emus enjoy playful interactions, such as chasing or being chased in a friendly manner. Ensure that any play activities are safe and avoid causing stress.

3. **Sensory Enrichment**

Stimulating an emu's senses can enhance their mental well-being:
- **Visual Stimuli:** Adding reflective objects or brightly colored toys in their environment can capture their attention and stimulate their curiosity.
- **Auditory Enrichment:** Emus respond to sounds, so playing soft music or introducing new sounds occasionally can engage their

auditory senses. However, avoid loud or sudden noises that may cause stress.
4. **Companionship and Social Interaction**
   Emus are social animals, and companionship plays a significant role in their emotional health:
   - **Other Emus or Livestock:** If possible, provide companionship in the form of other emus or compatible animals. Social interaction with others helps prevent loneliness and promotes natural behaviors.
   - **Human Interaction:** Regular, gentle interaction with you and other trusted humans provides emotional enrichment. Emus enjoy following their human companions around, which adds to their sense of security and happiness.

By providing a variety of enrichment activities that cater to their natural behaviors and interests, you can ensure that your emu remains mentally and physically stimulated. This leads to a healthier and more content bird, making your time together even more enjoyable.

Integrating emus into your daily life requires careful planning, mutual respect, and dedication. By understanding their unique needs and behaviors, building trust, and providing enriching activities, you can create a harmonious and fulfilling relationship with your emu. These fascinating birds will not only enrich your life but also become cherished companions when given the appropriate care and attention.

CHAPTER 7

BEYOND THE BASICS: ADVANCED TOPICS

Once you've mastered the essentials of emu care, training, and daily integration, there's a whole world of advanced topics to explore that can deepen your relationship with these majestic birds. Whether you're interested in advanced training, participating in emu shows and competitions, or learning about conservation efforts, this chapter will guide you through these exciting possibilities.

7.1 Advanced Training and Agility

While basic training helps establish a bond and manage emu behavior, advanced training techniques can further challenge your emu's mental and physical capabilities. This can also serve as an enjoyable way to deepen your interaction with your pet.

1. **Training for Agility**
    Emus are naturally agile, and harnessing this ability can be beneficial for their health and well-being. With structured agility training, you can tap into their innate capabilities to create an enriching, physically engaging experience for both the emu and yourself.
    - **Obstacle Courses:** Advanced obstacle courses, such as hurdles, tunnels, and balance beams, can be used to challenge the emu's coordination and speed. Start with simple obstacles and gradually increase the complexity as the emu becomes more confident.

- **Directional Commands:** Train your emu to respond to directional cues, such as "left," "right," "stop," and "go." This enhances their mental agility and helps build a deeper understanding of commands. Positive reinforcement, such as treats or affection, should be used to reward successful responses.
- **Recall and Obedience:** While emus are naturally independent, advanced recall training can help you teach your emu to come when called, follow you on walks, or respond to your commands. This level of control can be essential for safety, especially in larger, open spaces.

2. **Clicker Training**

Clicker training, a popular method for training many animals, can also be applied to emus. This method uses a small device that emits a distinct "click" sound, followed by a reward, to reinforce positive behaviors.
- **How Clicker Training Works:** Emus quickly learn to associate the click sound with a treat or other reward. You can use the clicker to reinforce behaviors such as walking through an obstacle course, standing still for grooming, or even performing tricks like bowing or stepping onto a platform.
- **Consistency and Patience:** Like with any animal, consistency is key when training emus. Patience is crucial, as emus may take longer to grasp certain concepts than smaller or more domesticated animals. Regular practice will yield positive results over time.

3. **Problem-Solving Games**
   Emus enjoy solving puzzles, and advanced training can involve the use of puzzle feeders, mazes, or interactive toys that require the emu to think critically to access rewards. Engaging them in problem-solving activities not only strengthens their cognitive skills but also serves as an excellent way to prevent boredom.

Advanced training takes emu care to the next level by providing physical and mental challenges. Through agility exercises, clicker training, and problem-solving activities, you can develop a closer bond while keeping your emu entertained and active.

7.2 Emu Shows and Competitions

For owners looking to showcase the beauty, agility, and intelligence of their emus, participating in shows and competitions can be an exciting and rewarding experience. While emu shows are less common than those for other livestock, they are growing in popularity, and the competitive aspect can serve as a motivating challenge for both owners and their birds.

1. **Preparing for Emu Shows**
   Participating in emu shows requires proper preparation, training, and grooming. Although emus are known for their rugged appearance, good grooming and presentation will help them stand out in competitions.
   - **Grooming and Presentation:** Regular grooming ensures that your emu's feathers are clean, glossy, and free from parasites. Special attention should be given to their plumage and nails before any show. While

emus are typically low-maintenance when it comes to grooming, a clean, well-presented bird is crucial for competition settings.
- **Training for Showmanship:** Emus that participate in shows must be trained to handle the presence of crowds, unfamiliar environments, and other animals. Begin with basic obedience training and work towards having your emu walk calmly on a lead, stand still when required, and respond to commands. Exposure to different environments during training will help your emu adapt to the show atmosphere.

2. **Types of Competitions**

Emu competitions can vary, ranging from physical contests to beauty shows. Some common categories include:
- **Conformation Shows:** These shows focus on the physical appearance and condition of the emu, with judges evaluating the bird based on breed standards, feather quality, and overall health. While emus don't have as many strict breed standards as other animals, judges still look for qualities like strong posture, vibrant plumage, and overall well-being.
- **Agility Contests:** In agility competitions, emus are judged on their ability to complete obstacle courses, follow commands, and showcase their physical prowess. These events highlight the bird's speed, agility, and coordination, allowing them to demonstrate their natural abilities.
- **Novelty Competitions:** Some emu shows may feature novelty events, such as trick contests or races. These lighthearted

Emu companion

competitions provide an opportunity to showcase unique behaviors and talents, adding a fun and entertaining element to the event.

3. **Benefits of Participation**
   Participating in emu shows and competitions can offer several benefits:
   - **Building Bond and Trust:** Preparing for competitions requires consistent training and interaction, which strengthens the bond between you and your emu.
   - **Showcasing Unique Talents:** Competitions allow emus to demonstrate their intelligence, agility, and personality in a public setting.
   - **Networking with Other Enthusiasts:** Emu shows are a great way to connect with other emu owners, exchange tips, and share experiences, helping to build a supportive community of enthusiasts.

Participating in emu shows and competitions requires preparation and dedication, but it's a rewarding way to engage with your emu and showcase their unique abilities.

7.3 Conservation Efforts and the Future of Emus

As a responsible emu owner, it is essential to understand the broader context of emu conservation and the role you can play in protecting these incredible birds. While emus are not currently endangered, their populations face threats due to habitat loss, climate change, and other environmental challenges.

1. **Conservation Status of Emus**
   The common emu, native to Australia, is classified as "Least Concern" by the International Union for

Conservation of Nature (IUCN), meaning that it is not currently at risk of extinction. However, several factors could potentially affect their populations in the future:
- **Habitat Destruction:** The primary threat to emus in the wild is habitat loss due to agricultural expansion, deforestation, and urban development. These factors limit the availability of natural foraging areas and disrupt breeding patterns.
- **Climate Change:** Changes in climate patterns can also affect emu populations by altering their food sources and migration patterns. Prolonged droughts, for example, can lead to food shortages and increased competition for resources.

2. **The Role of Captive Breeding Programs**
Captive breeding programs, often found in wildlife parks and conservation centers, play a vital role in preserving genetic diversity and ensuring that emu populations remain stable. These programs allow for controlled breeding in environments that protect emus from external threats such as habitat loss or predation.
- **Contributing to Conservation:** As an emu owner, you can contribute to conservation efforts by supporting captive breeding programs, donating to wildlife organizations, and spreading awareness about the importance of preserving emu habitats.

3. **The Future of Emu Ownership**
The future of emu ownership relies on responsible breeding practices, conservation awareness, and education. As emus become more popular as exotic pets and livestock, it is crucial that owners

understand their responsibilities in ensuring the welfare of their birds while contributing to conservation efforts.
- **Education and Advocacy:** Raising awareness about the ecological significance of emus and the importance of preserving their natural habitats can help protect wild populations. Educating others about the responsibilities of emu ownership, including proper care and ethical breeding practices, is essential in maintaining healthy populations both in captivity and in the wild.

Emu conservation efforts rely on the combined work of wildlife organizations, breeders, and responsible pet owners. By staying informed and supporting these initiatives, you can help ensure a sustainable future for emus in both the wild and captivity.

Chapter 7 delves into advanced topics of emu care and ownership, exploring everything from advanced training techniques to participation in emu shows and the larger context of emu conservation. These advanced aspects of emu ownership offer a deeper level of engagement with your bird and highlight the critical role owners can play in ensuring the future of these unique animals.

CHAPTER 8
HOUSING AND HOUSING SETUP FOR EMUS

Creating the right environment for your emu is crucial for their overall health, safety, and well-being. Proper housing not only ensures that your emu feels secure and comfortable but also provides a space that supports their natural behaviors, such as running, foraging, and exploring. In this section, we'll cover the essential elements of emu housing, from enclosure size to fencing, shelter requirements, and environmental enrichment.

1. Enclosure Size and Space Requirements

Emus are large, active birds that require plenty of space to roam, exercise, and exhibit natural behaviors. Providing a spacious enclosure is critical for both their physical and mental health. A cramped or undersized living area can lead to stress, aggression, and health issues.

- **Minimum Space Requirements:**
  The minimum recommended enclosure size for a single emu is at least **2500 square feet** (232 square meters). For a pair of emus, increase the size to at least **4000 square feet** (372 square meters). The larger the space, the better, as emus enjoy walking and running long distances. Having a spacious area allows them to stretch their legs and explore, which is important for preventing boredom and maintaining their physical health.
- **Exercise Area:**
  Emus need room to run, and their enclosure should include an open, flat space where they can move

freely. This is essential for preventing obesity and ensuring their muscles and joints remain strong. If possible, provide a separate, larger area for daily exercise outside of their primary living space.

- **Multiple Birds:**
  If you're housing more than one emu, ensure that the enclosure is large enough to allow each bird to have its own space. Emus are generally social animals, but overcrowding can lead to stress and aggressive behavior. Consider dividing the enclosure into sections if necessary, especially if you have multiple males, as they can become territorial.

2. Fencing and Security

Proper fencing is one of the most critical components of emu housing. Emus are strong, fast birds that can run at speeds of up to 30 miles per hour (48 kilometers per hour), so ensuring that your fencing is secure is essential to prevent escapes and keep your emu safe from predators.

- **Height of Fencing:**
  The fence surrounding your emu enclosure should be at least **6 feet (1.8 meters) high**. Emus are not known to jump or fly, but they can lunge or try to climb over lower fences if they feel threatened or if something outside the enclosure catches their attention.
- **Fence Material:**
  Choose a sturdy material for your fence, such as welded wire mesh or chain-link fencing. The gaps in the mesh should be small enough to prevent emus from getting their heads or legs caught. Avoid using barbed wire, as it can cause injuries to your bird.

- **Fence Durability:**
Since emus can push against fencing, ensure that the structure is well-secured and able to withstand their weight and strength. Fences should be supported by strong posts buried deep into the ground to prevent tipping or collapse, particularly if emus lean against them.
- **Predator Protection:**
If you live in an area where predators such as coyotes, dogs, or foxes are present, it's essential to make the enclosure predator-proof. Reinforce the bottom of the fence to prevent digging, and consider adding an overhead net or cover to protect your emu from aerial predators.

3. Shelter and Protection from the Elements

Emus are hardy birds and can adapt to a variety of climates, but they still require a shelter to protect them from extreme weather conditions, such as intense heat, rain, or cold temperatures. The shelter should provide a dry, well-ventilated space for the emu to rest, seek shade, and stay protected from the elements.

- **Size of Shelter:**
The shelter should be large enough for your emu to stand, lie down, and move around comfortably. For a single emu, a shelter of **10 x 10 feet (3 x 3 meters)** is adequate. If you are housing multiple emus, increase the size accordingly to ensure all birds can find cover inside without crowding.
- **Ventilation and Insulation:**
Good airflow is essential to prevent the buildup of moisture and ammonia from droppings, which can lead to respiratory issues. Ensure the shelter has ventilation openings, particularly at the top, to allow

for fresh air circulation. In colder climates, insulating the shelter can help keep the emu warm during freezing weather.

- **Flooring:**
Provide a dry, non-slip surface inside the shelter. Soft substrates like sand, straw, or wood shavings can be used to cushion the emu's feet and prevent injuries. Regularly check the floor for dampness and replace bedding materials to maintain a clean, dry environment.

- **Shade and Sun Protection:**
Emus are sensitive to heat, and in hot climates, shade is vital for preventing overheating. Provide additional shade structures in the enclosure, such as trees, awnings, or shade cloths, so that the emu can escape the sun during the hottest parts of the day.

4. Environmental Enrichment

Enrichment is critical to keeping your emu mentally and physically stimulated. A well-enriched environment can reduce boredom, stress, and destructive behaviors, while promoting natural instincts such as foraging and exploring.

- **Foraging Opportunities:**
Emus are natural foragers and will spend much of their time searching for food. Scatter food around the enclosure to encourage foraging, or use puzzle feeders to challenge them and keep them engaged. You can also plant native grasses and shrubs that emus enjoy grazing on.

- **Interactive Toys:**
Providing toys or objects for your emu to interact with can prevent boredom. Large balls, logs, or ropes can be placed in the enclosure for them to peck at or kick. You can also hang safe, sturdy toys

at various heights for them to investigate and push around.
- **Varied Terrain:**
Emus enjoy exploring different types of terrain. Incorporate different surfaces into the enclosure, such as gravel, sand, or dirt. This not only helps with foot health by providing different textures but also keeps the emu interested in its surroundings.
- **Water Features:**
Emus love water and will happily wade into ponds or splash in shallow pools. Adding a water feature or a large, shallow trough to the enclosure allows them to cool down, bathe, and engage in natural water play behaviors. Ensure that any water source is shallow and safe, with clean water available daily.

5. Regular Maintenance and Cleanliness

Maintaining a clean and hygienic environment is essential for preventing diseases and keeping your emu healthy. Regularly cleaning the enclosure, removing waste, and maintaining the quality of the shelter and fencing are necessary parts of responsible emu ownership.

- **Waste Management:**
Emus produce a significant amount of waste, and it's important to clean up droppings regularly to prevent the buildup of bacteria, parasites, and ammonia. Depending on the size of your enclosure and the number of birds, this may need to be done daily or several times per week.
- **Water Supply:**
Clean, fresh water should always be available. Emus require a large water container that is sturdy enough to prevent tipping. Change the water daily

to prevent contamination and ensure that it remains free from debris or algae buildup.
- **Check for Wear and Tear:**
Regularly inspect the fence, shelter, and other enclosure features for signs of wear, damage, or hazards. Replace or repair any damaged fencing immediately to ensure the safety of your emu.

Creating a well-designed, spacious, and secure living environment is fundamental to your emu's health and happiness. By providing ample space, secure fencing, weather-appropriate shelter, and enriching features, you can ensure that your emu thrives in a safe, stimulating, and comfortable habitat. Regular maintenance and attention to detail will help prevent potential health issues and allow your emu to live a full and content life in your care.

CHAPTER 9

GROOMING AND CARE

Emus are generally low-maintenance birds, but like all animals, they require regular grooming and care to stay healthy and comfortable. While their self-care instincts help with some aspects of their hygiene, your role as an owner is crucial in ensuring that they remain clean, free from parasites, and well-maintained. In this chapter, we will cover all the aspects of grooming and care, including feather maintenance, foot and nail care, and general hygiene practices to keep your emu in top condition.

1. Feather Maintenance and Grooming

Emus have unique feathers that serve several purposes, from protecting them against the elements to providing insulation. Although emus do not require the frequent grooming that some other birds or animals need, it's still important to monitor the condition of their feathers and provide assistance when necessary.

- **Natural Feather Care:**
  Emus generally take good care of their own feathers. They will often use dirt or sand baths to help keep their feathers free of parasites and debris. Providing a clean, dry area for them to dust bathe is essential for their self-maintenance. Dust baths help control mites and other parasites while also keeping their feathers from becoming matted with oils and dirt.
- **Periodic Feather Checks:**
  It's a good idea to periodically check your emu's feathers for signs of damage, mites, or other

Emu companion

parasites. Look for areas where feathers may be missing, dull, or overly greasy. This could indicate a problem that requires attention, such as an underlying health issue or a parasite infestation.

- **Assisted Feather Cleaning:**
While emus usually manage their feathers well, there are times when they may need help with cleaning, especially if they've been exposed to excessive mud or debris. Use a soft brush or cloth to gently remove any caked-on dirt. Avoid using water or soaps, as this can strip their feathers of natural oils. If your emu needs more intensive cleaning, consult a vet for advice on appropriate methods.
- **Molting:**
Like many birds, emus go through a molting process where they shed old feathers and grow new ones. This is a natural process that typically occurs once a year, usually in the warmer months. During molting, your emu may look a bit scruffy, and they may not be as active. Ensure that your emu has access to adequate nutrition during this time, as molting can put extra strain on their body.

2. Foot and Nail Care

Foot and nail care is a critical aspect of emu grooming. Emus have large, strong feet with tough claws that need to be monitored to prevent overgrowth or injuries.

- **Foot Health:**
Emus spend much of their time on their feet, and because of this, their foot health is of utmost importance. Check your emu's feet regularly for signs of injury, cuts, or abrasions. Common issues such as foot infections (bumblefoot) can develop if injuries go untreated. If you notice swelling,

redness, or the bird limping, consult a veterinarian immediately.
- **Nail Trimming:**
Emu nails can grow quite long, especially in captive environments where they may not wear down naturally as they would in the wild. If your emu's nails become too long, they can cause discomfort or even affect their ability to walk. Trim the nails as needed using a large nail clipper or grinder designed for large birds or livestock. Always take care not to cut too close to the quick (the blood vessel inside the nail), as this can cause pain and bleeding. If you're unsure about trimming your emu's nails, consult a vet or professional groomer who can show you the proper technique.
- **Providing Appropriate Substrate:**
To help prevent nail overgrowth and foot issues, ensure your emu has access to varied terrain. Surfaces like gravel, sand, and dirt can naturally wear down their nails and provide foot stimulation, reducing the need for frequent trims.

3. Beak and Eye Care

Although emus' beaks generally require little maintenance, it's essential to monitor their beak health to ensure they remain in good condition. Their eyes, too, can be prone to irritation or injury, so keeping an eye on these areas is important.

- **Beak Maintenance:**
Emu beaks naturally wear down as they peck at food and objects in their environment. However, if their beak becomes overgrown or damaged, it may require trimming. Signs of overgrowth include difficulty eating or picking up food. If you notice

Emu companion

these signs, consult a vet to ensure safe trimming. It's important to handle beak trimming carefully, as improper trimming can cause pain or long-term damage.
- **Eye Health:**
  Emus are curious animals that frequently investigate their surroundings. Their eyes can be prone to minor irritations from dust, dirt, or debris. Regularly check their eyes for redness, discharge, or swelling. If their eyes appear irritated, gently wipe away any debris with a clean, damp cloth. In cases of persistent irritation or injury, seek veterinary attention.

4. Parasite Control

Parasite control is a significant part of your emu's care routine. External parasites like mites, ticks, and lice can affect your emu's health and comfort, while internal parasites can lead to more serious health issues if left untreated.

- **External Parasites (Mites, Ticks, Lice):**
  Emus are susceptible to external parasites, which can cause itching, feather damage, and general discomfort. Regularly check your emu's skin and feathers for signs of mites or lice, which may appear as small moving dots or clusters near the base of feathers. Ticks can often be found in the folds of the skin around the neck and legs. To control external parasites, dusting powders or sprays designed for poultry and large birds can be used. Ensure the product is safe for emus, and consult a vet before applying any treatments.
- **Internal Parasites (Worms):**
  Internal parasites, such as worms, can also affect

emus. Symptoms of internal parasites include weight loss, lethargy, diarrhea, or a drop in appetite. Regular deworming is recommended to keep your emu free from these parasites. Deworming schedules vary depending on location and risk factors, so it's best to consult a vet to determine the appropriate deworming protocol for your emu.

- **Prevention and Cleanliness:**
  Maintaining a clean environment is one of the best ways to prevent parasite infestations. Regularly clean the enclosure, remove waste, and change bedding material. Keeping the living area dry and free from standing water also reduces the risk of parasites.

5. Bathing and Skin Care

While emus don't require frequent baths, they do benefit from occasional cleaning, especially if they've gotten particularly dirty. Bathing is also an opportunity to check for skin issues, such as dryness or irritation.

- **Dust Baths:**
  As mentioned earlier, emus often use dust baths to keep their skin and feathers clean. Providing a designated dust bath area with sand or fine dirt allows them to naturally care for their skin and remove parasites. This is an important part of their grooming behavior and should be encouraged.
- **Water Baths:**
  Emus enjoy water and may wade into shallow ponds or splash in water troughs. You can allow your emu to bathe in clean, shallow water, but avoid fully submerging them, as emus aren't aquatic animals. After bathing, ensure they have a dry,

shaded area where they can preen and dry off to avoid chills.
- **Skin Care:**
  While bathing, take the opportunity to inspect your emu's skin for dryness, irritation, or signs of injury. Dry patches or excessive flaking may indicate nutritional deficiencies or environmental issues. If you notice persistent skin problems, consult a vet to ensure there are no underlying health concerns.

6. Seasonal Care Adjustments

Emu care requirements can vary depending on the season. Being mindful of seasonal changes and how they affect your emu is essential for their comfort and well-being.

- **Summer Care:**
  During hot weather, it's important to ensure that your emu has access to plenty of shade and fresh, cool water. Emus can overheat easily, so providing a shaded shelter and water features, such as shallow pools, helps them stay cool. Regularly check for signs of heat stress, such as panting, lethargy, or open-mouth breathing.
- **Winter Care:**
  Emus are relatively hardy birds, but in colder climates, they may need additional care during winter. Ensure their shelter is insulated and protected from drafts, wind, and snow. Provide extra bedding for warmth, and monitor their water supply to prevent freezing. Cold weather can also exacerbate joint issues, so ensure they have access to dry, non-slippery surfaces.

Proper grooming and care are essential for maintaining your emu's health, comfort, and well-being. By staying vigilant and proactive about their feather, foot, and skin health, as well as keeping their living environment clean and free of parasites, you can ensure that your emu remains happy and healthy for years to come. While emus are relatively low-maintenance compared to some pets, regular attention to grooming and hygiene is key to preventing problems before they arise.

With this chapter, you are well-equipped to meet the grooming needs of your emu, helping them thrive under your care.

As you reach the end of this guide, you've not only gained practical knowledge on caring for emus, but you've also embarked on an extraordinary journey with one of the world's most unique and captivating animals. Whether you are a new emu owner, a seasoned enthusiast, or simply someone curious about these birds, the experience of caring for an emu can be deeply rewarding and transformative.

REFLECTING ON YOUR JOURNEY WITH EMUS

Owning an emu is unlike any other pet ownership experience. These magnificent creatures, with their prehistoric stature, inquisitive nature, and distinctive personalities, provide companionship and fascination that is hard to match. From the first moment you welcomed an emu into your life to the ongoing responsibilities of care, training, and enrichment, you have formed a bond that will continue to grow as long as you nurture it.

Throughout this journey, you've learned:

- **The Basics of Emu Care:** How to meet their dietary, housing, and health needs.
- **Training and Enrichment:** Techniques to engage their minds and bodies, ensuring they thrive in your care.
- **The Importance of Understanding Behavior:** How to communicate with your emu and foster a trusting relationship.
- **Advanced Opportunities:** Participating in emu shows, competitions, and exploring the role of emu conservation in preserving these extraordinary birds for future generations.

Reflecting on your progress, it's important to recognize the depth of your commitment. Emu ownership isn't just a hobby; it's a long-term responsibility that requires time, effort, and passion. As you continue to care for your emu, every step you take strengthens the bond between you and your bird, helping to ensure that they lead a healthy and fulfilling life.

Emu companion

Final Tips for Successful Emu Ownership

To wrap up this guide, let's revisit some final tips for making your journey with emus as successful as possible. These key points can serve as a reminder of what's most important in emu care:

1. **Consistency is Key:** Emus thrive in environments where they feel safe and secure. Establish a routine for feeding, exercise, and interaction, and stick to it. Consistency helps build trust and keeps stress levels low for your bird.
2. **Commit to Lifelong Learning:** The more you learn about your emu, the better equipped you will be to meet their needs. Don't be afraid to seek out new information, ask questions, and continually refine your approach to care.
3. **Stay Observant:** Emus are unique animals with specific behaviors and tendencies. Stay alert to changes in their behavior, eating habits, or physical condition, as these can be early indicators of health issues. Prompt attention to any signs of illness or discomfort will help you ensure their well-being.
4. **Provide Enrichment:** Keep your emu mentally stimulated with puzzles, toys, and activities that engage their natural instincts. A bored emu can become frustrated or stressed, so offering daily enrichment is essential to their happiness.
5. **Build Relationships:** Whether with other emus, livestock, or humans, emus are social creatures. Encourage healthy social interactions to keep them emotionally fulfilled, and continue nurturing your bond with regular interaction and affection.

6. **Health and Safety First:** Regular vet checkups, vaccinations, parasite control, and a clean environment are the foundations of long-term emu health. Don't neglect these essentials, and ensure that your emu's living space is safe, secure, and free from hazards.

By following these tips and continuing to dedicate yourself to your emu's well-being, you'll ensure that you're providing the best possible care. Successful emu ownership is about more than just meeting basic needs—it's about creating a life full of adventure, trust, and enrichment for your bird.

Resources and Further Reading

As you continue your journey with emus, it's essential to stay informed. The more knowledge you have, the better prepared you'll be for any challenges or changes that may arise in your emu's life. Below is a list of resources and further reading to support you as a dedicated emu owner:

1. **Books and Guides on Emu Care:**
    - *"The Complete Guide to Emu Farming"* by John K. Walker
    - *"Emu Health and Nutrition Handbook"* by Lisa M. Green
2. **Online Communities:**
    - **The Emu Owners Association:** A global community of emu owners sharing tips, experiences, and advice.
    - **Backyard Poultry Forums:** A great platform for those raising emus alongside other birds.
3. **Veterinary Resources:**

- **Association of Avian Veterinarians (AAV):** A comprehensive resource for finding veterinarians experienced in large bird care.
  - **The Emu Veterinary Guide:** A digital guide created by emu specialists that covers everything from vaccinations to advanced health concerns.
4. **Conservation and Breeding Organizations:**
   - **The Emu Conservation Network:** Dedicated to preserving emu populations in the wild and educating the public about the importance of conservation efforts.
   - **World Emu Federation:** Focused on promoting responsible breeding practices and the welfare of emus worldwide.
5. **Online Articles and Research Papers:**
   - **"Understanding Emu Behavior and Social Structures"** by Dr. James C. Turner
   - **"The Impact of Climate Change on Australian Emu Populations"** published in the *Journal of Avian Conservation*

By staying connected to these resources, you will always have access to new information and a community of fellow emu enthusiasts. Continued learning and engagement are crucial to ensuring your emu thrives in your care.

Final Thoughts

Owning an emu is a rare and extraordinary experience. It requires patience, dedication, and a willingness to learn, but the rewards are immense. As you continue on this journey, remember that emu care is about more than just meeting basic needs—it's about fostering a relationship based on mutual trust, respect, and understanding. With the

Emu companion

knowledge you've gained from this book, combined with your ongoing commitment, you are well-equipped to provide your emu with a life filled with happiness, health, and enrichment.

Enjoy your journey with these fascinating creatures, and may your relationship with your emu grow stronger each day. Whether as a companion, a show bird, or simply a fascinating pet, your emu will bring joy, wonder, and a deeper connection to the natural world into your life.

FREQUENTLY ASKED QUESTIONS (FAQS) ABOUT EMUS AS PETS

General Questions

1. **Are emus good pets for families?**
    - Emus can be great pets for families with ample space and a secure outdoor area. They are generally friendly and curious but may not be suitable for families with very young children due to their size and strength.
2. **How long do emus live?**
    - Emus typically live between 10 to 20 years in captivity, with proper care and management.
3. **Can emus be kept indoors?**
    - Emus are large birds that require significant outdoor space to roam. They are not suited for indoor living, as they need room to exercise and display natural behaviors.

Legal Considerations

4. **Do I need a permit to own an emu?**
    - Ownership regulations for emus vary by location. In some areas, a permit or license is required, while others may have restrictions on ownership. Always check local laws and regulations before acquiring an emu.
5. **Are there any specific zoning laws for keeping emus?**
    - Yes, zoning laws differ between municipalities. Ensure your property complies with local zoning regulations

Emu companion

concerning livestock and exotic animals before getting an emu.

Care and Housing

6. **What type of enclosure do emus need?**
    - Emus need a secure outdoor enclosure with sturdy fencing (at least 6 feet tall) to prevent escape and protect them from predators. The enclosure should include ample space to roam, forage, and exercise.
7. **What should I feed my emu?**
    - Emus require a balanced diet consisting of commercial emu feed, grains, vegetables, and fruits. Fresh water should always be available. Consult with a vet for specific dietary recommendations based on age and health.
8. **How much space do emus need?**
    - Emus require a minimum of one acre of space per bird. The more space you provide, the happier and healthier your emus will be.

Health and Grooming

9. **What are common health issues in emus?**
    - Common health issues include bumblefoot (foot infections), respiratory infections, and gastrointestinal problems. Regular veterinary check-ups are essential to monitor your emu's health.
10. **How often should I check my emu for parasites?**
    - Regular checks (at least once a month) for external parasites such as mites, ticks, and lice are recommended. Consult a vet for

Emu companion

advice on deworming and parasite prevention.

11. **Do emus require special grooming?**
    - Emus don't require extensive grooming, but regular checks for feather health, nail trimming, and foot care are essential. They benefit from dust baths for feather maintenance.

Behavior and Training

12. **Are emus social animals?**
    - Yes, emus are social creatures that thrive in groups. Keeping them in pairs or small groups is recommended to prevent loneliness and stress.
13. **Can emus be trained?**
    - Emus can be trained using positive reinforcement techniques. Basic commands and tricks can be taught, but patience and consistency are key.
14. **What should I do if my emu shows aggressive behavior?**
    - Aggressive behavior in emus can stem from fear or territorial instincts. Ensure they have a safe space, provide enrichment, and consult a veterinarian or animal behaviorist if aggressive behaviors persist.

Breeding and Reproduction

15. **How do emus breed?**
    - Emus reach sexual maturity around 2-3 years of age. Breeding involves pairing males and females during the breeding season, typically from late spring to

Emu companion

summer. Males incubate the eggs after they are laid.

16. **What is involved in caring for emu chicks?**
    - Emu chicks require a warm and safe environment, high-quality starter feed, and constant access to fresh water. They need protection from predators and must be monitored for health issues as they grow.

Miscellaneous

17. **Can emus be kept with other animals?**
    - Emus can coexist with other farm animals, but care should be taken to monitor interactions, as their size and behavior may intimidate smaller animals. Always supervise mixed-species interactions.
18. **What should I do if I need to travel and can't take my emu with me?**
    - If you need to travel, consider arranging for a trusted friend, neighbor, or professional caretaker to look after your emu. Ensure they understand emu care needs.
19. **Where can I find a vet that specializes in exotic birds or emus?**
    - Look for veterinarians specializing in avian or exotic animal care in your area. Online directories, local animal hospitals, or recommendations from emu owners can be valuable resources.
20. **What should I do if my emu escapes?**
    - If your emu escapes, remain calm. Try to corral them back to safety using treats or their favorite food. If you cannot recapture them quickly, contact local animal control for assistance.

Emu companion

www.ingramcontent.com/pod-product-compliance
Lightning Source LLC
Chambersburg PA
CBHW070211230526
45471CB00002B/922